Understanding the Elements of the Periodic Table™

THE ALKALI METALS

Lithium, Sodium, Potassium, Rubidium, Cesium, Francium

Kristi Lew

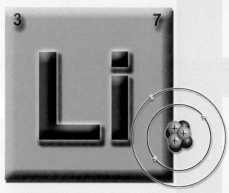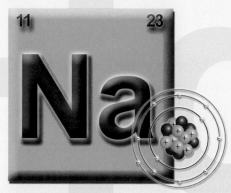

rosen publishing's
rosen central®

New York

Published in 2010 by The Rosen Publishing Group, Inc.
29 East 21st Street, New York, NY 10010

Library of Congress Cataloging-in-Publication Data

Lew, Kristi.
The alkali metals: lithium, sodium, potassium, rubidium, cesium, francium / Kristi Lew. — 1st ed.
    p. cm. — (Understanding the elements of the periodic table)
Includes bibliographical references and index.
ISBN-13: 978-1-4358-5330-0 (library binding)
1. Alkali metals—Popular works. 2. Periodic law—Popular works.
I. Title.
QD172.A4L49 2010
546'.38—dc22

                                                                    2009002245

Manufactured in the United States of America

On the cover: The elements lithium (Li), sodium (Na), potassium (K), rubidium (Rb), cesium (Cs), and francium (Fr) make up the alkali metals.

Contents

Introduction

The alkali metals are a group of six very useful elements—lithium (chemical symbol: Li), sodium (Na), potassium (K), rubidium (Rb), cesium (Cs), and francium (Fr). In fact, you could not live without the elements sodium and potassium. Your body needs them to function correctly. Rubidium, cesium, and francium are harder to find in everyday life, but lithium is very common. If you have ever used a small, portable electronic device like a laptop computer, an MP3 player, or a calculator, for example, it probably had a battery in it that contained the element lithium.

The alkali metals are found on the far left-hand side of the periodic table in group 1 (or IA in an older naming system). The periodic table of elements is a chart that chemists use to organize all of the known chemical elements. One of the first periodic tables was developed by a Russian chemist named Dmitry Mendeleyev (also spelled Dmitri Mendeleev; 1834–1907) in 1869.

Mendeleyev's table contained all sixty-three chemical elements known to scientists at that time arranged by increasing atomic weight. Each column contained elements with similar physical and chemical properties. With the elements arranged in this manner, their properties exhibited a periodic, or regular, pattern. When the known elements did

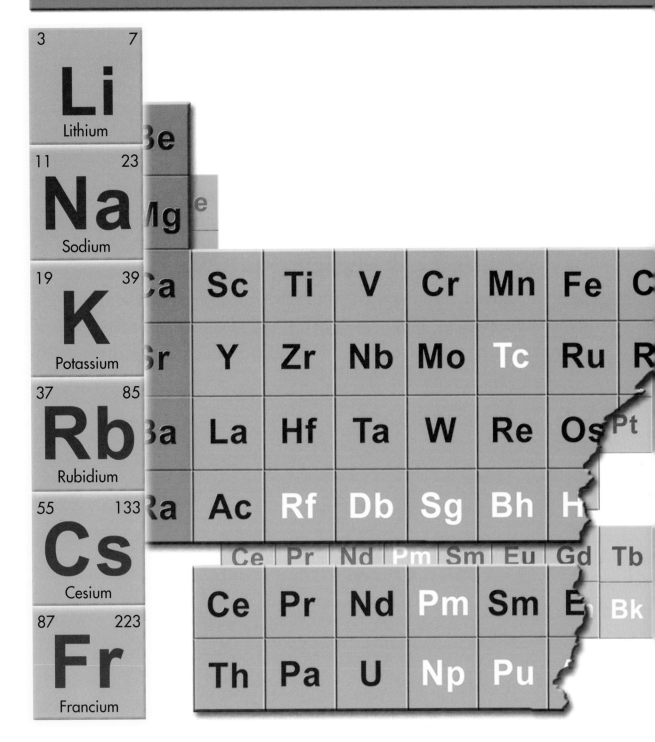

The six alkali metals can be found on the far left-hand side of the periodic table in group 1 or IA.

not fit into this regular pattern, Mendeleyev left a blank. He predicted that, one day, scientists would find elements with properties to fill in the blanks on his periodic table. Mendeleyev was proven correct when the elements gallium (Ga), scandium (Sc), and germanium (Ge) were discovered between 1875 and 1886. Each of these elements exhibited properties predicted by Mendeleyev and fit into one of the blank spots.

The modern periodic table is still based on Mendeleyev's table. However, the elements are no longer arranged by increasing atomic weight. Instead, the modern periodic table is arranged according to increasing atomic number. The rows on the periodic table are called periods. Lithium is in period 2. The other alkali metals are in periods three through seven. The columns of the periodic table are called groups or families. The alkali metals are in group 1, which is called the alkali metal group.

Chapter One
The Alkali Metals

Like all elements in the same group, the alkali metals have similar physical and chemical properties. For example, the group 1 elements are all metals. Metals are found on the left-hand side and in the middle of the periodic table. Metals, including the alkali metals, have similar physical properties. Physical properties are characteristics that can be measured or observed without changing the substance. Color, luster (shininess), and hardness are all physical properties. For example, the alkali metals all have low melting and boiling points. And, like most metals, they are malleable and ductile. Malleable means that they can be hammered into different shapes without breaking. Ductile is a similar property that means that they can be drawn

The density of lithium is so low that it floats in the mineral oil it is stored in.

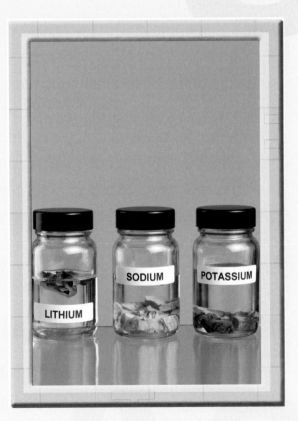

LITHIUM SODIUM POTASSIUM

The alkali metals react violently with water. For this reason, they are usually stored under oil.

into thin wires. The alkali elements also have very low densities, especially for metals. Objects with a low density are light for their size. In fact, lithium, sodium, and potassium have such low densities that they would float on water.

However, all the alkali metals have a common chemical property that makes putting them into water a very bad idea. When alkali metals come into contact with water, they can explode and catch on fire.

Chemical properties describe how substances change when they come in contact with other substances. The fact that the alkali metals react violently with water is a chemical property of these metals. The reaction causes the metals to change chemically. Of all the metals on the periodic table, the alkali metals are the most reactive. In fact, they react so easily that they are never found in nature in their pure form. Instead, the metals are chemically bonded with nonmetals into chemical compounds. A chemical compound is made up of two or more elements combined in a fixed ratio that is a property of the compound.

The History of the Alkali Metals

In 1807, Sir Humphry Davy (1778–1829), a British chemist, became the first person to isolate pure sodium metal. Sodium is the sixth-most-common

Sir Humphry Davy isolated the elements sodium, potassium, and lithium from their natural chemical compounds.

element on Earth, but it is always found chemically bonded to other elements in a chemical compound. The most common source of sodium is a mineral called halite, or rock salt. A mineral is a naturally occurring crystalline (made up of crystals) substance. Salt, diamond, and quartz are all common minerals. The mineral halite is almost pure sodium chloride ($NaCl$). This is the same chemical compound that makes up ordinary table salt. Sodium chloride can also be obtained by evaporating the water from seawater.

The element was named for one of its compounds, sodium carbonate ($NaCO_3$), which was known long before the element sodium was identified. The common name for sodium carbonate is soda ash, and the compound was often used to cure headaches. The word "sodium" comes from the English word "soda." Its strange chemical symbol, Na, comes from the Latin word *natrium*.

The same year that Davy isolated sodium, he also discovered the element potassium. Ranked seventh most abundant in Earth's crust, potassium is just slightly less common than sodium. The element is found in many minerals, including the mineral sylvite. The chemical name for sylvite is potassium chloride (KCl). To isolate the element, Davy used another potassium compound called potassium hydroxide (KOH). The element was named for the compound potassium carbonate, which has the common name potash. Like sodium, potassium's strange chemical symbol comes from Latin. The medieval Latin word for alkali is *kalium*.

Lithium was discovered by Johan August Arfvedson (1792–1841) in 1817. He found the element in a mineral called petalite [$LiAl(Si_2O5)_2$]. Lithium was named for the Greek word for stone.

Cesium was discovered by Robert Bunsen (1811–1899) and Gustav Kirchhoff (1824–1887) in 1860. They discovered it using a technique known as spectrometry. To use this technique, scientists heat an element until it glows. They look at the light produced by the glowing element through a prism. When sunlight is looked at through a prism, the prism breaks

Boom!

When the alkali metals encounter water, the chemical reaction that occurs releases very flammable hydrogen gas. Sometimes, the heat from the chemical reaction is enough to make the gas catch on fire. As you move down the group, the alkali metals get more and more reactive. For example, when lithium, the least reactive of the alkali metals, is put into water, it usually just bubbles like soda water as the reaction releases hydrogen gas. When sodium or potassium is dropped into water, however, they react so strongly that they burst into flames. And cesium reacts so violently that the shockwave it produces can shatter glass. Because of their violent reactions to water, the alkali metals are usually stored under mineral oil to prevent water in the atmosphere from reaching them.

the sunlight into all the colors of the rainbow. However, light given off by an individual element does not produce all of these colors. Instead, each element produces a unique set of bright lines in just a few colors. These lines are called spectral lines.

This is one way scientists can tell one element from another. When Bunsen and Kirchhoff looked at the spectral lines given off by cesium, they found that some of the element's lines were bright blue. The scientists named the element for the Latin word *caesius*, which means "sky blue." Small amounts of cesium can be found naturally, mostly in a mineral called pollucite ($Cs_4Al_4Si_9O_{26}$). Large amounts of this mineral are mined at Bernic Lake, in Manitoba, one of the Canadian provinces.

The year after discovering cesium, Bunsen and Kirchhoff also discovered rubidium. When they isolated the element, it produced the deepest red spectral lines they had ever seen. Therefore, the element was named

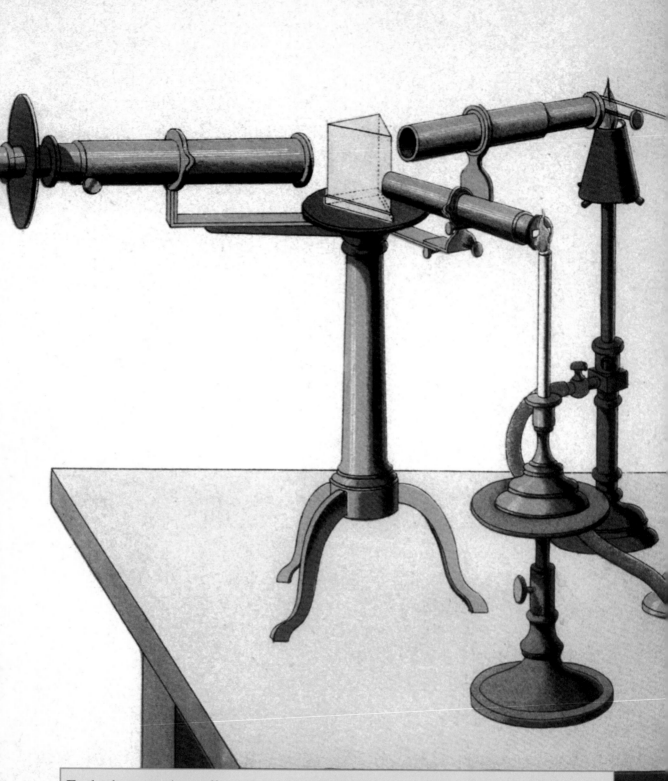

Each element gives off a unique set of spectral lines when heated to glowing and viewed through the prism of a spectrometer.

rubidium for the Latin word that means a deep red color. Rubidium is relatively abundant in Earth's crust. It is not as common as sodium or potassium, but it is about as common as copper (Cu), zinc (Zn), and nickel (Ni). The element can also be found in seawater and mineral springs.

Francium was the last of the alkali metals to be discovered. It was found by Marguerite Catherine Perey (1909–1975), a French chemist, in 1939. Francium is very rare on Earth. It is also a radioactive element. Radioactive elements are not stable, and over time, they break down into different elements. This process is called radioactive decay. Francium is produced by the radioactive decay of another element called actinium (Ac). Because francium also goes through radioactive decay and becomes a completely different element, scientists think that francium might be the rarest element in Earth's crust. In fact, they believe that only about 25 grams (less than one ounce) of francium exist at any one time. Francium was named for Perey's home country.

Chapter Two
Atomic Structure of the Alkali Metals

All elements are forms of matter. Matter is anything that takes up space and has weight. All matter—the air we breathe, the water we drink, the food we eat, and everything else, including you— is made up of tiny, invisible building blocks called atoms.

Subatomic Particles

Atoms are made up of even smaller objects called subatomic particles. There are three main subatomic particles—the electron, the proton, and the neutron. The English physicist Joseph John Thomson (1856–1940) discovered the electron in 1897. Electrons are negatively charged particles that move around the nucleus, or the central core, of an atom in spaces called energy levels, or shells.

After Thomson discovered the electron, scientists began to suspect that there were an equal number of positively charged particles in an atom. In 1911, Ernest Rutherford (1871–1937) discovered that nearly all of the weight of an atom was concentrated in a very tiny center that has a positive charge. Rutherford had discovered the nucleus. By the 1920s, scientists were calling the positively charged particles in the nucleus "protons." Rutherford is generally given credit for their discovery as well.

Joseph John Thomson and Ernest Rutherford helped scientists understand the makeup of atoms.

The number of protons in an element's nucleus gives that element its identity. In other words, each element has a particular number of protons. The number of protons in an atom of an element is called the element's atomic number. For example, lithium has an atomic number of three. Therefore, a lithium atom also has three protons. The number of protons in an atom is also equal to the number of electrons in that atom.

In 1913, Henry Moseley (1887–1915), a British chemist, suggested that the physical and chemical properties of the elements on the periodic table actually depended on the element's atomic number instead of atomic weight. He was basically correct. The chemical properties of an element depend largely on the arrangement of that element's electrons in their energy levels. Because the number of electrons is equal to the number of protons and the number of protons is equal to the atomic number, the properties of elements do repeat, in a periodic manner, according to increasing atomic number. Chemists still use a periodic table arranged in this manner today.

The third subatomic particle, the neutron, was not discovered until 1932. The neutron was discovered by Sir James Chadwick (1891–1974),

Hydrogen

On most periodic tables, hydrogen (H) is listed at the top of the alkali metal group. However, hydrogen is not considered an alkali metal. It is a nonmetal. However, hydrogen and the alkali metals do have something in common. Hydrogen's atomic number is one. Therefore, hydrogen has only one electron. That electron is on hydrogen's one and only energy level, which is also its highest energy level. Therefore, hydrogen, like the alkali metals, has one valence electron.

a British scientist. Unlike electrons and protons, neutrons were more difficult to detect because they have no electric charge. They are neutral. Like protons, neutrons are found in the nucleus. Together, protons and neutrons make up the majority of an atom's weight.

Atomic Structure of the Alkali Metals

Starting at the top of group 1 with the element lithium, the alkali metals' atomic numbers range from three to eighty-seven (for francium). A neutral lithium atom has three protons and three electrons. On the periodic table, an element's row, or period, indicates how many energy levels are occupied by the electrons in its atoms. Lithium is in period 2. Therefore, the element's three electrons are found on two energy levels. The first energy level can hold two electrons. The last electron is found on the second, and outermost, energy level. Electrons on the outermost energy level of an atom are called valence electrons. Lithium has one valence electron. Valence electrons are the electrons that are involved in chemical bonding with other atoms.

3	7

Li

11	23

Na

19	39

K

37	85

Rb

55	133

Cs

87	223

Fr

The alkali elements all chemically react in a similar manner because they all have one valence electron.

The next element in the alkali metal group is sodium. Sodium's atomic number is 11. Therefore, every sodium atom must have eleven protons and eleven electrons. Sodium is in period 3. Its eleven electrons can be found moving around its nucleus on three energy levels. The first energy level can hold two electrons. The second energy level can hold eight. That leaves one electron left over for the third, and outermost, energy level. Therefore, sodium, like lithium, has one valence electron. In fact, all of the alkali metals have only one valence electron. Because electrons are the subatomic particles that participate in chemical reactions, the fact that all of the alkali metals have one valence electron makes these elements react in a similar manner. In other words, because they all have the same number of valence electrons, they all have similar chemical properties.

Radioactivity

Most atoms have at least as many neutrons as protons. Neutrons help to keep the positively charged protons from repelling one another and breaking the nucleus apart. This is what makes some elements radioactive. As these elements break down, or decay, they give off energy called radiation. Radioactive elements give off radiation and continue to decay until their nuclei find a stable state. Because the number of protons changes as the nucleus decays, these elements change into entirely different elements as they decay. Actinium, for example, decays into the element francium. In turn, francium decays into the element astatine (At). The time it takes for half of a radioactive element to decay into another element is called its half-life.

The periodic table can also be used to determine the number of neutrons in an average atom of any element. Neutrons and protons have the same approximate weight—1.6726×10^{-27} kg. Because this number is so small, it is difficult to work with in calculations. To solve this problem, chemists decided that 1.66×10^{-27} kg would be equal to one atomic mass unit (amu). Therefore, a proton has a weight of 1 amu and so does a neutron. On the periodic table, the approximate weight of an average atom of each element is listed. For example, the approximate weight of an average lithium atom is 7 amu. Because the weight of one proton is 1 amu and lithium has three protons, the protons in a lithium atom weigh 3 amu. Neutrons make up the other 4 amu of the atom's weight. Therefore, an average lithium atom has four neutrons.

Chapter Three
Uses for the Alkali Metals

When elements chemically react with one another, their valence electrons are exchanged or shared. This exchange or sharing creates chemical bonds.

Ionic Bonding

When metals create a chemical bond with nonmetals, they lose their valence electrons. When an atom loses or gains electrons, its charge is no longer neutral. Chemists call these charged particles ions. When a metal loses its valence electrons, it has more protons than electrons. The extra protons give a metal ion an overall positive charge. The alkali metals react very easily because they have only one valence electron to lose.

Nonmetal atoms, on the other hand, gain electrons during chemical bonding. The extra electrons give a nonmetal ion a negative charge. When metals and nonmetals chemically react with one another, the opposite charges of their ions hold the ions together in a chemical bond. This type of chemical bond is called an ionic bond.

Chemical compounds that are held together by ionic bonds are called ionic compounds. For example, sodium chloride is a very common ionic compound. It is made up of a positively charged sodium ion and a

Sodium chloride, or table salt, is a very common ionic compound. Chemists refer to all ionic compounds as salts.

negatively charged chloride ion. The compound is commonly called salt, but chemists define a salt differently. In chemistry, a salt is any ionic compound made up of positive and negative ions. Therefore, technically, all ionic compounds are salts.

Getting Pure Alkali Metals

Many of the alkali metals are obtained in their pure form by using a technique called electrolysis. Electrolysis is the process of passing electricity through a solution or molten liquid that contains ions. To isolate potassium

Sir Humphry Davy was the first chemist to use electrolysis to isolate pure alkali metals from their dissolved or melted compounds. The same process is still used today.

for the first time, for example, Sir Humphry Davy used this technique on molten potassium hydroxide (KOH). The liquid contains positive potassium ions and negative hydroxide ions.

When electricity flows through a liquid that contains ions, chemical changes occur. At the positive electrode, electrons are pulled off of the negatively charged hydroxide ions. When electrons are stripped off of hydroxide ions, the ions decompose into oxygen gas and water. At the negative electrode, an electron is added to a potassium ion. With the addition of that one electron, the potassium again has the same number of protons and neutrons and is a neutral atom.

Metallic Bonds

Many of the physical properties exhibited by pure metals can be explained by the way their valence electrons behave. In a metal, these electrons are not tightly bound to any one particular atom. Instead, they are free to float between the atoms in the metal. To picture this, imagine a pan filled first with marbles and then with water. The marbles represent the nuclei and all of the metal's electrons, except for its valence electrons. These are fixed in place and don't move. Its valence electrons are represented by the water that is free to flow around and between the marbles.

The free-floating electrons allow metals to bend without breaking. If someone hammers on a metal, for example, its valence electrons just move out of the way, allowing the metal to bend. These free-floating electrons can also "carry" heat and electricity from one spot in a metal to another. This property allows metals to conduct heat and electricity well.

Today, the most abundant source of potassium is a mineral called sylvinite (KClNaCl). In the United States, a sylvinite mine in Carlsbad, New Mexico, produces about 85 percent of the potassium needed to make fertilizer. Fertilizers are the most common use of potassium compounds. Like pure sodium, pure metal potassium has limited uses.

Davy also used electrolysis to isolate sodium metal from one of its compounds, sodium hydroxide (NaOH). However, he was unable to obtain pure sodium by dissolving the compound in water. The solution did not work because as soon as pure sodium formed, it reacted with the water to form sodium hydroxide again. However, melting a compound will also break the compound into its ions. When Davy passed electricity through molten (melted) sodium hydroxide, he was successful in producing pure sodium. Today, this same method is sometimes used with the compound sodium chloride to produce pure sodium metal. However, there are only a few uses for pure sodium metal.

Today, pure lithium is also obtained by electrolysis of melted lithium chloride (LiCl). Lithium chloride can be obtained by allowing seawater to evaporate. When the water is gone, it leaves behind solid salts including lithium chloride, sodium chloride, and potassium chloride (KCl).

Rubidium and cesium metal are also produced by passing electricity through their molten compounds. Francium is made through the natural radioactive decay of the element actinium. It can also be produced artificially in the laboratory.

Airplanes and Batteries

Lithium is a very lightweight metal. Because of its weight, it is often alloyed with other metals to form lightweight materials for aircraft. An alloy is a mixture of two or more elements, one of which is a metal. Often, alloys are produced by melting two or more metals together and allowing them to

To save weight, the space shuttle's external fuel tanks are made from an alloy of aluminum and lithium.

cool. Brass, bronze, and steel are three common alloys. Brass is a mixture of copper and zinc. Copper is alloyed with tin (Sn) to produce bronze. And steel is a mixture of iron (Fe) and carbon (C) (a nonmetal). Alloys are often used instead of pure metals to capitalize on the different properties of the elements in the alloy. To make lighter aircraft parts, for example, lightweight lithium is usually alloyed with stronger metals, such as aluminum (Al) or magnesium (Mg). A mixture of these metals makes light, but strong, metal parts.

Lithium is also a metal that is used extensively in batteries. A battery is a device that converts chemical energy into electrical energy. A car battery, for example, uses the chemical reaction between lead (Pb) and lead dioxide (PbO_2) to supply a car with the needed electricity to run the radio, headlights, and interior lights. Lithium, however, is much lighter than lead. Therefore, lighter batteries that are suitable for use in portable electronics, such as laptop computers, calculators, watches, and MP3 players, can be made from lithium.

Streetlights and Electricity

Because sodium and potassium are very reactive elements, they are not used very often in their pure form. The most common use for pure sodium is in streetlights. Sodium metal that is heated to a very high temperature turns into a gas, or vapor. Sodium vapor is used in streetlights to produce a very bright characteristic yellow light.

Pure sodium and potassium are sometimes used in their liquid forms as a coolant, or heat exchange material, in nuclear reactors. The reactions that occur in nuclear reactors produce large amounts of heat. Coolants are used to transport that heat away from the reactor and to a boiler. Here, the heat causes water to boil, producing steam. The steam is used to turn the large, rotating blades of a machine called a turbine. The motion of the turbine is used to produce electricity.

Sodium vapor streetlights, like the ones shown here, give off a yellowish glow.

Uses for the Other Alkali Metals

The other alkali metals are used even less often in their pure form. Some forms of rubidium and cesium can be used to make atomic clocks. These clocks measure time very precisely, but because they use radioactive forms of rubidium and cesium, they are not for ordinary use.

A more common use of the two elements is in the manufacture of photocells. Photocells convert light energy into electrical energy. When sunlight, for example, hits a photocell, the light excites, or energizes, electrons. Energized electrons can break away from their atoms. This produces a flow of electrons. A flow of electrons, or any charged particles, produces electric current. Photocells are often used in devices such as alarm systems. A device in the alarm system is used to produce a beam of light. The light beam hits a photocell and produces an electric current. If someone passes through the beam of light, the electric current stops. This "trips" the alarm, making it sound and alerting the alarm owner that someone has "broken" the beam of light.

Pure sodium, potassium, rubidium, and cesium may not have many practical uses, but francium has even fewer. Not only is the element quite rare in nature, but only small amounts of it have ever been made in the laboratory. Because it decays so quickly (its half-life is about twenty-two minutes), there are currently no uses for francium outside of basic scientific research.

Chapter Four
The Alkali Metals in Compounds

There may not be many uses for the alkali metals in their pure form, but life would be extremely different without all of the alkali metal compounds that are used in everyday life.

Lithium Compounds

Lithium carbonate (Li_2CO_3) is a lithium compound that is added to glass or ceramic to make such material stronger. For example, the temperature-resistant, glass cookware with the trade name Pyrex is made out of glass that contains this lithium compound. Another place where strong glass comes in handy is in the lenses and mirrors of large telescopes, like the 200-inch (508 centimeters) mirror on the Hale Telescope at the Palomar Observatory in San Diego County, California. A mirror this large must be made out of strong glass so that it does not crack or distort the images astronomers are taking of the skies.

Lithium carbonate is also used in extracting aluminum metal from aluminum ores. When an aluminum ore is mined, it contains many impurities. To remove these unwanted substances, the aluminum ore must be heated to a very high temperature. However, if lithium carbonate is added, the same reaction can take place at a much lower temperature. Because the reaction takes place at a lower temperature,

The addition of the compound lithium carbonate makes the glass on the Hale Telescope's 200-inch mirror stronger. Scientists must clean the mirror before it gets a new coating of aluminum.

aluminum produced using lithium carbonate costs less to purify because less energy is needed.

Another compound of lithium, lithium stearate ($LiC_{18}H_{35}O_2$), is mixed with petroleum to make a heavy, lubricating grease used in many industrial machines. The grease works well because it does not break down at high temperatures and yet it does not get hard at low temperatures. It also does not react with water or oxygen in the air, so the chemical composition of the grease does not change over time and it continues to do its job.

Lithium hydroxide (LiOH) is another useful lithium compound. This compound can be used to absorb carbon dioxide. Too much carbon dioxide in small, enclosed spaces with limited oxygen supply, such as spacecraft or submarines, can cause dizziness, headaches, unconsciousness, and even death. Scientists are also concerned that too much carbon dioxide in the atmosphere may be causing the average global temperature to rise. Lithium hydroxide is one chemical that industrial manufacturing plants can use to scrub the carbon dioxide out of the air they release, making it cleaner and better for the environment.

Sodium Compounds

Sodium compounds are even more plentiful and useful than those of lithium. Ordinary table salt, baking soda, baking powder, drain cleaners, aspirin, soaps, and many other products that people use every day all contain one sodium compound or another.

The one people are most familiar with is salt. Salt, or sodium chloride, has been used for many centuries to flavor and preserve food. Sodium bicarbonate ($NaHCO_3$), or baking soda, is also used when baking. When this chemical compound comes into contact with an acidic compound, such as milk or vinegar, the resulting chemical reaction releases carbon dioxide gas. This gas forms bubbles, causing dough to "rise," making breads and cakes fluffy instead of flat. Some medications, such as Alka-Seltzer, also make use of this bubbling reaction to help settle the stomach. Mouthwashes, cleaning products, and fire extinguishers often contain sodium bicarbonate as well.

Sodium hydroxide ($NaOH$), which is also called lye or caustic soda, is a common ingredient in oven and drain cleaners. The compound can also be used to make soap. Sodium hydroxide is not the only sodium compound to have many names. Sodium carbonate (Na_2CO_3) is also called soda, soda ash, or washing soda. One of this compound's uses is the manufacture of glass. Glass is made by heating sodium carbonate and calcium oxide, also called lime, and sand together. When the mixture

Sodium compounds, such as baking soda, baking powder, and salt, are needed to make freshly baked pretzel treats.

cools, it turns hard and transparent. Sodium carbonate is also used in the manufacturing of textiles, paper, and soaps and detergents.

Potassium Compounds

The most common use for potassium compounds is in fertilizers because plants need potassium to grow. Potassium chloride (KCl) and potassium nitrate (KNO_3), or saltpeter, are the most common potassium compounds used for this purpose. Potassium chloride is sometimes used as a salt substitute, too. Potassium nitrate is also used to make match heads and pyrotechnics, or fireworks. Two other potassium compounds, potassium chlorate ($KClO_3$) and potassium perchlorate ($KClO_4$), are also used in explosives and fireworks. Potassium hydroxide (KOH) is useful in the manufacturing of soaps and detergents. And, like sodium hydroxide, it is often used in drain cleaners. Potassium tartrate ($KC_4H_5O_6$), also called cream of tartar, is a common ingredient in baking. And potassium carbonate (K_2CO_3) is used to make glass and soaps.

Other Alkali Metal Compounds

Rubidium makes many compounds. However, most of them have no commercial purposes.

Potassium compounds are most commonly used in fertilizers.

The Alkali Salts and Fireworks

If you have ever seen a fireworks display during a Fourth of July celebration, you have seen one of the more entertaining uses for the alkali salts. Many alkali compounds are used to make the colors in fireworks. Lithium salts, for example, are used to give fireworks a red color. Sodium salts produce yellow fireworks. Potassium salts are used in fireworks, too. But they do not give the fireworks color themselves. Instead, they provide the fuel that produces the heat needed to burn the other metal salts and produce color.

Many alkali salts are used to make colorful fireworks.

One possible exception is a compound made up of rubidium, silver (Ag), and iodine (I) ($RbAg_4I_5$), which scientists think may one day be used to make thin film batteries.

Cesium has a few useful compounds. Cesium chloride (CsCl) and cesium nitrate ($CsNO_3$) are two of the most common. Both of these compounds are used to produce other useful chemicals. Cesium chloride and cesium carbonate (Cs_2CO_3) are sometimes used to make beer. And a radioactive form of cesium chloride is used to treat certain types of cancer.

Because francium is so rare and it decays so quickly, there are no useful francium compounds.

Chapter Five
The Alkali Metals and You

Many of the alkali metal compounds are useful in everyday life. From soap to table salt and fertilizers to drain cleaners, life would be very different without these compounds.

The Alkali Metals in the Human Body

Sodium chloride is used to add flavor to food. But that is not the only reason it is added. Salt also preserves food. It keeps bacteria from growing in it so that the food does not spoil. This method of preserving food has been used for many centuries. Even today, many canned and prepared

The right balance of sodium and potassium is necessary for a healthy body.

foods have a lot of sodium chloride added to them. This concerns many heath experts because many people eat more than twice the amount of sodium chloride than their bodies need. In humans, sodium helps regulate the amount of fluid inside cells. Too much or too little fluid can prevent cells from functioning correctly. And too much salt can lead to a dangerous condition called hypertension, or high blood pressure. People with high blood pressure have a higher risk of suffering serious health problems such as strokes or heart attacks. Some people use a salt substitute instead of regular table salt to avoid these problems. Many of these substitutes are made up of potassium chloride instead of sodium chloride.

Potassium is also an important element for the body. Like sodium, it helps regulate the amount of fluid in cells and, in many cases, works with sodium to make them function properly. Potassium is also an important element for plants. In fact, they cannot live without it. And because all plants contain potassium, humans have no problem getting the amount of potassium needed in their diets, even without salt substitutes.

Corrosive Chemicals

The hydroxides of sodium, potassium, and cesium are extremely caustic, or corrosive, chemicals. Corrosive chemicals can do irreparable harm or even destroy objects in which they come into contact. All of these chemicals can cause severe burns on the skin, eyes, and respiratory passages. Sodium and potassium hydroxides are often used in drain and oven cleaners because they can easily dissolve fats. However, these chemicals can also dissolve the fats that make up the human body if they are not used correctly. Cesium hydroxide ($CsOH$) is the most caustic of all the alkali metal hydroxides. In fact, it is corrosive enough to dissolve glass.

Bipolar Disorder

In 1949, an Australian physician named John Cade (1912–1980) found that people with bipolar disorder responded favorably to taking the compound lithium carbonate. People with bipolar disorder often experience extreme mood swings, from very up and frenzied (a condition doctors call manic) to very down and depressed. Because of these symptoms, this condition is also sometimes called manic-depressive disorder.

Lithium carbonate was the first drug therapy found to help the disorder. Doctors do not know exactly how the drug helps to stabilize a person's mood, but they do know that it works on the central nervous system (brain and spinal cord). Today, more than 60 percent of people who suffer from bipolar disorder take lithium to help control their symptoms. Studies have shown that the drug significantly lowers the suicide risk in people with the disorder and generally makes the "highs" not quite so high and the "lows" not quite as low.

The amount of lithium in the body needs to be closely monitored by a doctor, however. Too much lithium can lead to unwanted side effects such as uncontrollable shaking in the hands, increased thirst and urination, weight gain, trouble with memory and concentration, and

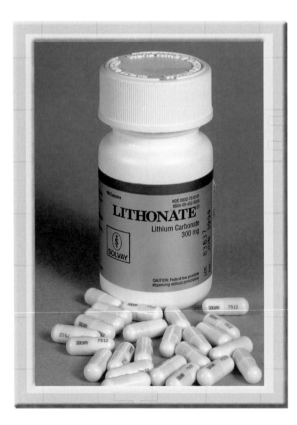

Lithium carbonate is one medicine that may be used to control the symptoms of bipolar disorder.

Soap

Sodium hydroxide, which is also called lye, and potassium hydroxide are often used to make soap. When a strong base (the chemical opposite of an acid), like either of these two chemicals, is heated with a fat or a vegetable oil, soap is formed. This process is called saponification.

extreme sleepiness. The drug can also affect the kidneys, and a prolonged imbalance can lead to kidney damage. With a doctor's help, however, the dose of lithium carbonate can be adjusted to relieve most of the side effects and still reduce the symptoms of bipolar disorder.

Because the alkali metals, and especially their compounds, are so common, everyday life would be very different without them. Without baking soda, salt, and salt substitutes, our food would not taste quite the same. We could not clean our houses or ourselves without soaps, detergents, and drain cleaners. We also rely on glass, paper, batteries, and matches made from alkali metal compounds. Not to mention the fact that our bodies cannot function without the elements sodium and potassium. Indeed, our very lives depend on the alkali metals and their compounds.

The Periodic Table of Elements

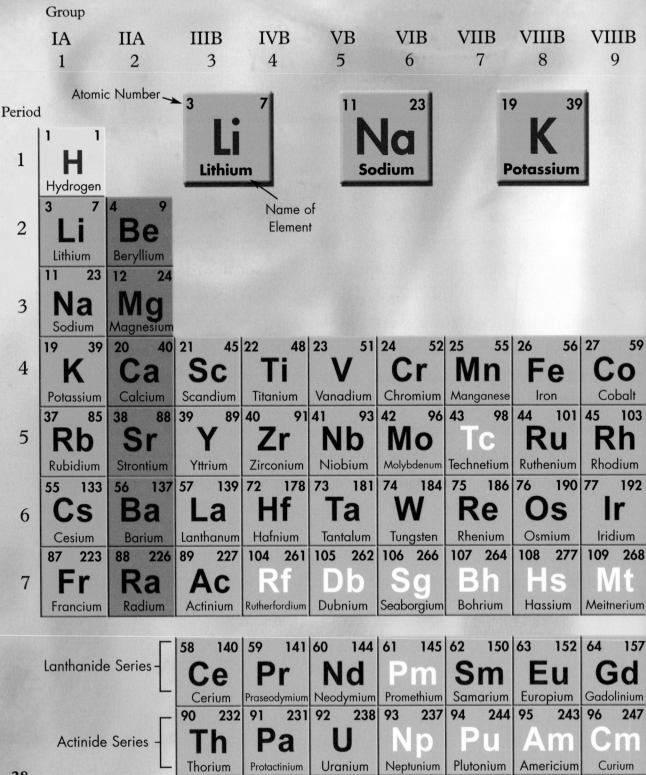

Group

IA	IIA	IIIB	IVB	VB	VIB	VIIB	VIIIB	VIIIB
1	2	3	4	5	6	7	8	9

Atomic Number →

3   **7**
Li
Lithium

Name of Element

11   **23**
Na
Sodium

19   **39**
K
Potassium

Period

1   **1**
H
Hydrogen

Period								

Period 2

3 7 **Li** Lithium	4 9 **Be** Beryllium

Period 3

11 23 **Na** Sodium	12 24 **Mg** Magnesium

Period 4

19 39 **K** Potassium	20 40 **Ca** Calcium	21 45 **Sc** Scandium	22 48 **Ti** Titanium	23 51 **V** Vanadium	24 52 **Cr** Chromium	25 55 **Mn** Manganese	26 56 **Fe** Iron	27 59 **Co** Cobalt

Period 5

37 85 **Rb** Rubidium	38 88 **Sr** Strontium	39 89 **Y** Yttrium	40 91 **Zr** Zirconium	41 93 **Nb** Niobium	42 96 **Mo** Molybdenum	43 98 **Tc** Technetium	44 101 **Ru** Ruthenium	45 103 **Rh** Rhodium

Period 6

55 133 **Cs** Cesium	56 137 **Ba** Barium	57 139 **La** Lanthanum	72 178 **Hf** Hafnium	73 181 **Ta** Tantalum	74 184 **W** Tungsten	75 186 **Re** Rhenium	76 190 **Os** Osmium	77 192 **Ir** Iridium

Period 7

87 223 **Fr** Francium	88 226 **Ra** Radium	89 227 **Ac** Actinium	104 261 **Rf** Rutherfordium	105 262 **Db** Dubnium	106 266 **Sg** Seaborgium	107 264 **Bh** Bohrium	108 277 **Hs** Hassium	109 268 **Mt** Meitnerium

Lanthanide Series

58 140 **Ce** Cerium	59 141 **Pr** Praseodymium	60 144 **Nd** Neodymium	61 145 **Pm** Promethium	62 150 **Sm** Samarium	63 152 **Eu** Europium	64 157 **Gd** Gadolinium

Actinide Series

90 232 **Th** Thorium	91 231 **Pa** Protactinium	92 238 **U** Uranium	93 237 **Np** Neptunium	94 244 **Pu** Plutonium	95 243 **Am** Americium	96 247 **Cm** Curium

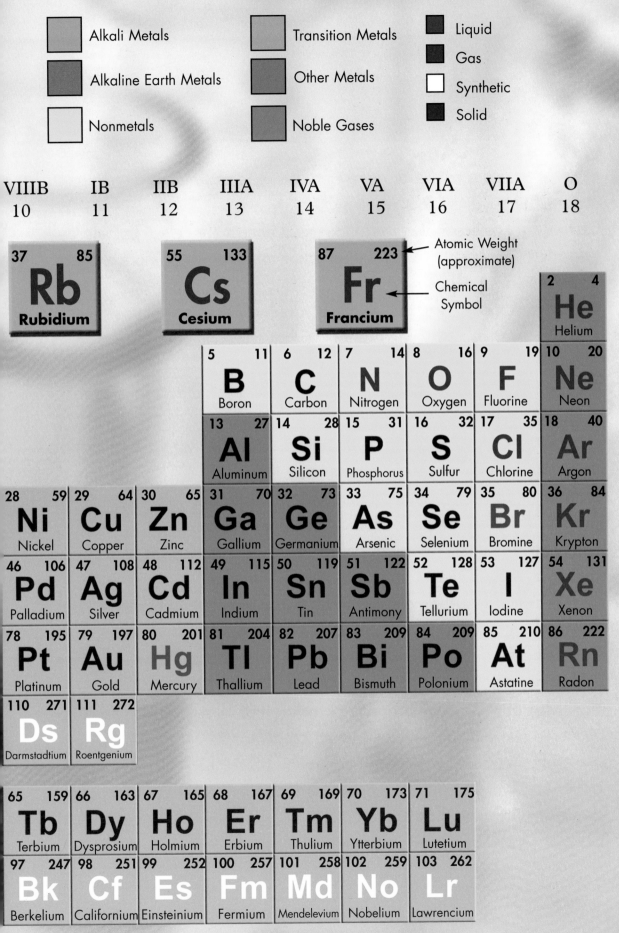

Legend

- Alkali Metals
- Alkaline Earth Metals
- Nonmetals
- Transition Metals
- Other Metals
- Noble Gases
- Liquid
- Gas
- Synthetic
- Solid

VIIIB 10	IB 11	IIB 12	IIIA 13	IVA 14	VA 15	VIA 16	VIIA 17	O 18

37 85 Rb Rubidium
55 133 Cs Cesium
87 223 Fr Francium — Atomic Weight (approximate) — Chemical Symbol

2 4 He Helium

5 11 B Boron
6 12 C Carbon
7 14 N Nitrogen
8 16 O Oxygen
9 19 F Fluorine
10 20 Ne Neon

13 27 Al Aluminum
14 28 Si Silicon
15 31 P Phosphorus
16 32 S Sulfur
17 35 Cl Chlorine
18 40 Ar Argon

28 59 Ni Nickel
29 64 Cu Copper
30 65 Zn Zinc
31 70 Ga Gallium
32 73 Ge Germanium
33 75 As Arsenic
34 79 Se Selenium
35 80 Br Bromine
36 84 Kr Krypton

46 106 Pd Palladium
47 108 Ag Silver
48 112 Cd Cadmium
49 115 In Indium
50 119 Sn Tin
51 122 Sb Antimony
52 128 Te Tellurium
53 127 I Iodine
54 131 Xe Xenon

78 195 Pt Platinum
79 197 Au Gold
80 201 Hg Mercury
81 204 Tl Thallium
82 207 Pb Lead
83 209 Bi Bismuth
84 209 Po Polonium
85 210 At Astatine
86 222 Rn Radon

110 271 Ds Darmstadtium
111 272 Rg Roentgenium

65 159 Tb Terbium
66 163 Dy Dysprosium
67 165 Ho Holmium
68 167 Er Erbium
69 169 Tm Thulium
70 173 Yb Ytterbium
71 175 Lu Lutetium

97 247 Bk Berkelium
98 251 Cf Californium
99 252 Es Einsteinium
100 257 Fm Fermium
101 258 Md Mendelevium
102 259 No Nobelium
103 262 Lr Lawrencium

Glossary

alloy A mixture of two or more elements, one of which is a metal.

atom The building block of all matter.

battery A device that converts chemical energy into electrical energy.

chemical compound A substance that contains two or more elements in a fixed ratio.

density A measure of the weight of an object per unit volume.

ductile A physical property of metals that allows them to be drawn into thin wires.

electrolysis The process of passing electricity through a solution to obtain pure elements.

electron A negatively charged subatomic particle that travels around the outside of the nucleus in an energy level.

family A column on the periodic table of elements.

group A column on the periodic table of elements.

half-life The time it takes for half of a radioactive element to decay into another element.

ion A charged particle created when an atom or group of atoms loses or gains electrons.

ionic bond A chemical bond created by the attraction between positive and negative ions.

malleable A physical property of metals that allows them to be hammered into different shapes without breaking.

mineral A naturally occurring crystalline substance.

neutron A neutral subatomic particle found in the nucleus of an atom.

nucleus The central core of an atom.

period A row on the periodic table of elements.

physical property A property of a substance that can be measured or observed without changing the substance.

proton A positively charged subatomic particle found in the nucleus of an atom.

radioactive decay The process that unstable elements undergo to become different elements.

radioactive element An unstable element that breaks down into another element over time.

salt An ionic compound made up of a positive ion and a negative ion.

spectral lines The unique set of bright colored lines given off by each element when it is heated to glowing and viewed through a prism.

spectrometry The science of using spectral lines to tell what elements make up a substance.

valence electron An electron on the highest energy level that is involved in chemical bonding.

For More Information

American Chemical Society
1155 Sixteenth Street NW
Washington, DC 20036
(800) 227-5558
Web site: http://www.acs.org
The American Chemical Society provides many educational resources, including experiments to try, information about National Chemistry Week, and a fun periodic table called the periodic table of elephants.

Chemical Institute of Canada
130 Slater Street, Suite 550
Ottawa, ON K1P 6E2
Canada
(613) 232-6252
Web site: http://www.cheminst.ca
The Chemical Institute of Canada provides information about available science fairs, scholarships, and the Canadian Chemistry Contest.

National Institutes of Health
9000 Rockville Pike
Bethesda, MD 20892
(301) 496-4000
Web site: http://www.nih.gov
The National Institutes of Health can answer all types of health questions, including those about the recommended amount of sodium and potassium in your diet.

The Salt Institute
700 N. Fairfax Street, Suite 600
Fairfax Plaza
Alexandria, VA 22314-2040
(703) 549-4648
Web site: http://www.saltinstitute.org
The Salt Institute answers frequently asked questions about salt and the
salt industry.

Soap and Detergent Association
1500 K Street NW, Suite 300
Washington, DC 20005
(202) 347-2900
Web site: http://www.cleaning101.com
The Soap and Detergent Association provides information about soaps
and detergents, including what they are made up of and how to
make your own.

Web Sites

Due to the changing nature of Internet links, Rosen Publishing has developed an online list of Web sites related to the subject of this book. This site is updated regularly. Please use this link to access the list:

http://www.rosenlinks.com/uept/tam

For Further Reading

Brent, Lynnette. *Acids and Bases*. New York, NY: Crabtree Publishing Company, 2008.

Browning, Marie. *Totally Cool Soapmaking for Kids*. New York, NY: Sterling Publishing, 2005.

Cooper, Sharon Katz. *The Periodic Table: Mapping the Elements*. Mankato, MN: Capstone Press, Inc., 2007.

Jerome, Kate Boehm. *Atomic Universe: The Quest to Discover Radioactivity*. Des Moines, IA: National Geographic Society Children's Books, 2006.

Masters, Nancy Robinson. *Salt*. Ann Arbor, MI: Cherry Lake Publishing, 2008.

Roza, Greg. *Potassium*. New York, NY: Rosen Publishing Group, 2007.

Silverstein, Alvin. *Depression and Bipolar Disorder Update*. Berkeley Heights, NJ: Enslow Publishers, Inc., 2008.

Slade, Suzanne. *Elements and the Periodic Table*. New York, NY: Rosen Publishing Group, 2006.

Tocci, Salvatore. *Sodium*. New York, NY: Children's Press, 2006.

Walker, Denise. *Acids and Alkalis*. Mankato, MN: Black Rabbit Books, 2007.

Bibliography

AbsoluteAstronomy.com. "Carbon Dioxide Scrubber." Retrieved January 1, 2009 (http://www.absoluteastronomy.com/topics/Carbon_dioxide_scrubber).

American Cancer Society. "Cesium Chloride." Retrieved January 1, 2009 (http://www.cancer.org/docroot/ETO/content/ETO_5_3X_Cesium_Chloride.asp).

Chemical Heritage Foundation. "Julius Lothar Meyer and Dmitri Ivanovich Mendeleev." Retrieved January 1, 2009 (http://chemheritage.org/classroom/chemach/periodic/meyer-mendeleev.html).

ENotes. "Alloy: World of Earth Science." Retrieved January 1, 2009 (http://www.enotes.com/earth-science/alloy).

Gray, Theodore. "Alkali Metal Bangs." Retrieved January 1, 2009 (http://www.theodoregray.com/PeriodicTable/AlkaliBangs).

Jefferson Lab. "It's Elemental—The Periodic Table of Elements." Retrieved January 1, 2009 (http://education.jlab.org/itselemental/index.html).

Oracle ThinkQuest Library. "CHEMystery: Electrochemistry: Electrolysis." Retrieved January 1, 2009 (http://library.thinkquest.org/3659/electrochem/electrolysis.html).

Royal Society of Chemistry. "Visual Elements: Group 1—The Alkali Metals." Retrieved January 1, 2009 (http://www.rsc.org/chemsoc/visualelements/Pages/data/intro_groupi_data.html).

Sawyer, Lee. "Who Discovered the Proton?" Retrieved January 1, 2009 (http://www.physlink.com/Education/askexperts/ae46.cfm).

Soap and Detergent Association. "Soap Chemistry." Retrieved January 1, 2009 (http://www.cleaning101.com/cleaning/chemistry).

Stwertka, Albert. *A Guide to the Elements.* 2nd ed. New York, NY: Oxford University Press, 2002.

Thomson Gale. "Chemical Elements." Retrieved January 1, 2009 (http://www.chemistryexplained.com/elements/index.html).

WebMD. "Bipolar Disorder: Lithium Treatment." Retrieved January 1, 2009 (http://www.webmd.com/bipolar-disorder/ bipolar-disorder-lithium).

Index

About the Author

Kristi Lew is the author of more than twenty science books for teachers and young people. Fascinated with science from a young age, she studied biochemistry and genetics in college. Before she started writing full-time, she worked in genetics laboratories for more than ten years and taught high school science. When she's not writing, she enjoys sailing with her husband aboard their small sailboat, Proton. She writes, lives, and sails in St. Petersburg, Florida.

Photo Credits

Cover, pp. 1, 5, 17, 38–39 by Tahara Anderson; p. 7 © Martyn F. Chillmaid/Photo Researchers, Inc.; p. 8 © Charles D. Winters/Photo Researchers, Inc.; p. 9 Private Collection/The Bridgeman Art Library; p. 12 © World History/Topham/The Image Works; p. 15 © Bettmann/Corbis; p. 21 © Yaov Levy/Phototake; p. 22 © SSPL/The Image Works; p. 25 NASA Kennedy Space Center; p. 26 © A. T. Willett/Alamy; p. 29 © Roger Ressmeyer/Corbis; p. 31 © www.istockphoto.com/Benjamin Christie; p. 32 © AP Images; p. 33 © www.istockphoto.com/Lyle Gregg; p. 34 © www.istockphoto.com/RiverNorthPhotography; p. 36 © Phil Degginger/Alamy.

Designer: Tahara Anderson; Editor: Bethany Bryan;
Photo Researcher: Amy Feinberg